SPOTLIGHT ON PHYSICAL SCIENCE

ENERGY

SO MUCH POTENTIAL

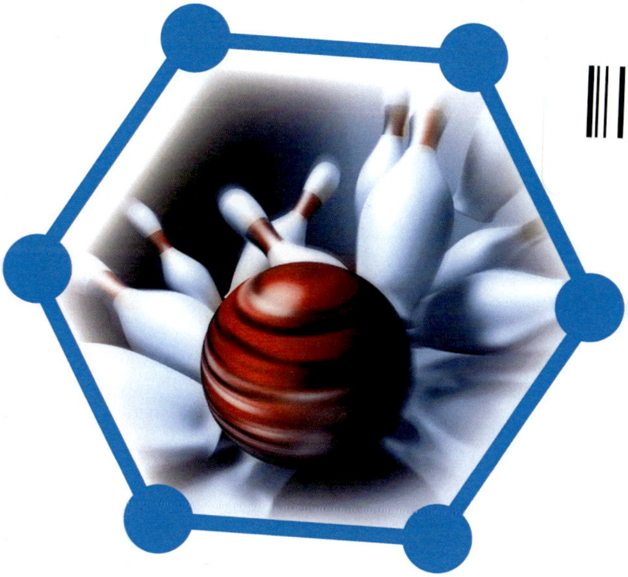

LISA IDZIKOWSKI

PowerKiDS press

NEW YORK

Published in 2020 by The Rosen Publishing Group, Inc.
29 East 21st Street, New York, NY 10010

Editor: Jennifer Lombardo
Book Design: Michael Flynn

Photo Credits: Cover, p. 1 bioraven/Shutterstock.com; (series molecular background) pro500/Shutterstock.com; p. 5 Alberto Masnovo/Shutterstock.com; p. 6 Vadym Lesyk/Shutterstock.com; p. 7 Everett Historical/Shutterstock.com; p. 8 Oleksiy Mark/Shutterstock.com; p. 9 grafxart/Shutterstock.com; p. 11 (top) lightpoet/Shutterstock.com; p. 11 (bottom) David Carillet/Shutterstock.com; p. 12 Smileus/Shutterstock.com; p. 13 SlayStorm/Shutterstock.com; p. 14 WDG Photo/Shutterstock.com; p. 15 canadastock/Shutterstock.com; p. 17 -/AFP/Getty Images; p. 18 Fer Gregory/Shutterstock.com; p. 19 Suzi44/Shutterstock.com; p. 21 Chrispo/Shutterstock.com; p. 23 Leonard Zhukovsky/Shutterstock.com; p. 24 Ger Bosma Photos/Shutterstock.com; p. 25 Chris Holman/Shutterstock.com; p. 27 Larina Marina/Shutterstock.com; p. 29 mandritoiu/Shutterstock.com.

Cataloging-in-Publication Data

Names: Idzikowski, Lisa.
Title: Energy: so much potential / Lisa Idzikowski.
Description: New York : PowerKids Press, 2020. | Series: Spotlight on physical science | Includes glossary and index.
Identifiers: ISBN 9781725313040 (pbk.) | ISBN 9781725313071 (library bound) | ISBN 9781725313057 (6 pack)
Subjects: LCSH: Energy industries--Juvenile literature. | Biomass energy--Juvenile literature. | Renewable energy sources--Juvenile literature.
Classification: LCC TJ163.23 I49 2020 | DDC 621.042--dc23

Manufactured in the United States of America

CPSIA Compliance Information: Batch #CWPK20. For further information contact Rosen Publishing, New York, New York at 1-800-237-9932.

CONTENTS

CHAPTER ONE

ENERGY EVERYWHERE

Computers running and lights flickering, friends playing video games or kicking a soccer ball, parents cooking dinner: all of these things take energy. Energy makes things happen, causes changes, and gets work done.

There are two main types of energy: potential energy and kinetic energy. Potential energy is the energy stored in objects that aren't moving. Kinetic energy is the energy of motion or moving objects. Every action we take uses one of these types of energy.

Energy is also the way we make the things around us work. Without it, we wouldn't have cars, electric lights, cell phones, or a lot of other things we use every day. Some types of energy are **renewable**, so they can be easily used over and over. Others are nonrenewable and can only be used once. Energy can't be created or destroyed, only changed into different forms. We can't see or touch energy, but it's always around.

There are many kinds of energy, and it can come from many places.

CHAPTER TWO

WHEN DID IT ALL BEGIN?

Early humans used fire for energy. Burning wood gives off light and heat that they could use to stay warm and cook their food. At first, most people lived in small groups, but when these groups started to turn into larger towns and cities, they needed more energy.

Steam produced with water and burning coal provided energy that powered many different machines, including trains and boats.

People learned how to use the energy from many different sources. They learned that water and wind could be used to produce energy to make machines run. By the mid-1700s, people found that burning coal, a **fossil fuel**, could power more complicated machines. Engineers, scientists, and inventors got busy! Creative people came up with many different ways to use the energy sources they'd discovered. Over time, they created all sorts of machines—big and small, simple and complicated—that people use every day.

CHAPTER THREE

AT THE SOURCE

Prehistoric people didn't use a lot of energy except what they needed to stay alive. They got this energy from the food they ate, which powered their bodies. Today, though, many more things, including cars, phones, and computers, need power. There are many different ways people create the large amount of energy used around the world every day.

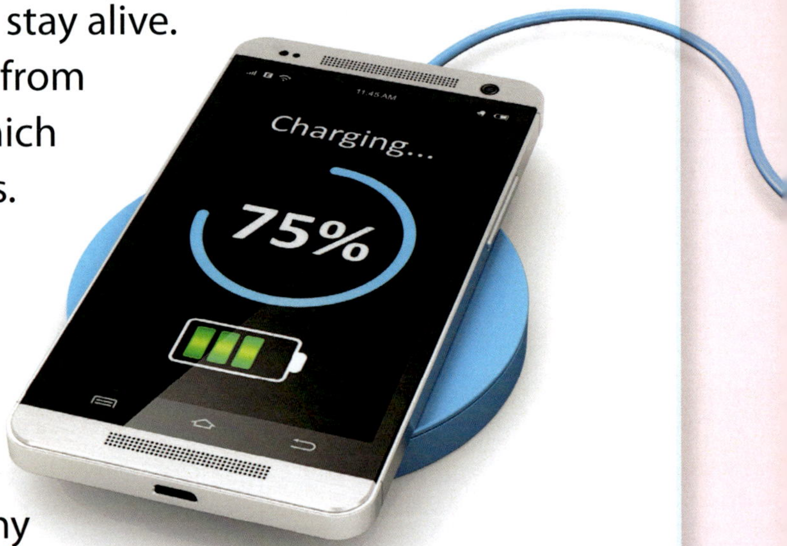

In the United States, most energy comes from fossil fuels. These are nonrenewable sources of energy, so they'll be all used up one day. This is one reason scientists are working to find better ways to use renewable energy sources. Power that comes from the sun, wind, water, and other renewable sources, such as **biomass**, won't run out if the sources are used properly.

Biomass can be a source of renewable energy. Plant materials, such as the trees shown here, can be used to create energy.

CHAPTER FOUR

ALWAYS CHANGING

Energy is everywhere, and the amount is always the same; you can't make more and you can't get rid of what already exists. Instead, energy is always changing from one form to another.

If energy couldn't change, you couldn't hop on a bike to ride to a friend's house or pull on skates and glide across the ice. When you eat food, energy from the food is stored in your muscles as potential energy. Then, when you're ready to ride, skate, play soccer, or do anything else, the potential energy turns into kinetic energy as your muscles move.

Machines work because of energy transformations, too. Many people today use cell phones. **Chemical energy** from a cell phone's battery changes into electrical energy, which is what makes the phone work.

This picture shows the difference between potential energy and kinetic energy.

POTENTIAL ENERGY

KINETIC ENERGY

CHAPTER FIVE

THE BIGGEST ENERGY SOURCE

It feels good to be outside on a sunny day, but the sun does more than just light up the sky and make the air warm. In fact, the sun is Earth's main source of energy.

When you eat plants, you're taking in the energy they get from the sun.

This may seem odd, since a lot of the energy we use every day comes from burning fossil fuels. What's the connection? Fossil fuels such as coal, oil, and natural gas formed from dead plants and animals over millions of years. When they were alive, these plants made and stored energy through **photosynthesis**. The animals got their energy by eating the plants or by eating other animals that had eaten plants.

Earth is exactly the right distance from the sun for life to exist. If it were a little bit closer, it would be too hot and dry here for anything to live; a little farther away would make it too cold and dark.

CHAPTER SIX

WIND AND WATER

Wind and water are two renewable sources of energy. We use wind energy harnessed by big machines called **turbines**. When the wind blows, it moves the blades of the turbine. As those turn, they move a different machine inside the turbine called a generator. The generator changes the kinetic energy of the wind to electrical energy. When the wind stops blowing, the generator stops making electricity.

Water can also be used for electrical power. Power plants that use water energy to make electricity are called hydroelectric plants. As fast-moving water goes through a water turbine, it turns a generator inside the turbine. There are other ways to use waterpower, too. People can build dams to make **reservoirs**. When electricity is needed, workers at the dam can release the water from the reservoir quickly so it makes the generator move.

Lake Mead, a reservoir created by Hoover Dam, is the largest reservoir in the United States in terms of volume.

CHAPTER SEVEN

FLIP THE SWITCH

Many people use a lot of electrical power today. When it gets dark, we turn on the lights. When it gets too cold, we turn on the heat, which may be powered by electricity or natural gas. When we want to relax, we turn on the television or computer. So many things we use in everyday life depend on the electrical energy we harness and use as electricity.

People didn't always have electricity in their homes. Instead, they used candles and oil lamps to light their houses. Iceboxes kept food cold, and burning wood or coal helped houses stay warm.

Many scientists and inventors throughout history experimented with electricity. Today, electrical engineers design systems to send electricity everywhere it's needed.

Thomas Edison didn't invent the first light bulb, but he did make light bulbs easier and safer for people to use in their houses.

CHAPTER EIGHT

ENERGIZED ANIMALS

People and animals harness energy in many ways. Humans find ways to produce light energy with many different tools for many reasons. Some animals make their own light with their own bodies. This is called bioluminescence. In this process, they turn chemical energy into light energy. A chemical in their body called luciferin reacts with oxygen, making parts of the animal's body light up. Some animals that have bioluminescence can control when they glow. Some can even change the color of their lights!

This drawing of an anglerfish shows its bioluminescent lure.

Many people have seen fireflies on summer nights, but most bioluminescent creatures live deep in the ocean. One example is the anglerfish. This kind of fish has a bulb on the top of its head that lights up. This tricks smaller fish into coming close enough for the anglerfish to eat them.

CHAPTER NINE

WAY TOO HOT!

Everything around us is made up of atoms and molecules of matter. These particles are constantly vibrating, or moving back and forth very fast. The faster they vibrate, the more energy they have. Thermal energy, which we often call heat, is the energy an object has because of this particle motion. Temperature is the way we measure thermal energy.

Thermal energy can be used many different ways. Imagine a bonfire. The heat energy from the fire can travel through the air to warm you up or perfectly toast your marshmallows. Toasters use heat energy to turn bread into toast. Another type of thermal energy, geothermal energy, comes from deep within Earth. Engineers have created ways to bring geothermal energy to the surface through pipes. This natural energy can be used to heat buildings.

Geothermal power plants use large pipes such as the ones shown here to get heat from the plant to people's houses.

CHAPTER TEN

USING GRAVITY FOR ENERGY

Imagine being in a pool on a hot day. You're holding a bucket full of water above your head, ready to let it splash down. Now think about a tall waterfall. Water at the top falls a long way to the bottom—much farther than water from the bucket. Both examples show a type of energy. Gravitational potential energy is an object's potential energy that has to do with gravity, the force that pulls something toward the center of Earth.

Another example of something showing gravitational energy is a snowboarder. As a snowboarder starts their run, gravity pulls them down the hill. Speeding down, they go faster and faster. When they hit a bump, they fly so high into the air because of this energy that they have time to perform fancy tricks before they hit the ground!

Shaun White is a world-class snowboarder. Here, he's shown competing in the men's snowboard half-pipe event at the 2018 Winter Olympics.

CHAPTER ELEVEN

SOUND ENERGY

Sound is a form of energy, too. Singing birds and chirping crickets rely on sound energy to get their messages out to the world. People can hear these sounds of nature because sound travels through the air in waves to our ears. If there were no air around us, we wouldn't be able to hear sounds. This is why you wouldn't be able to hear anything in outer space!

Sound travels faster in liquid than it does through the air. When whales call to each other underwater, they can hear each other sooner than they would be able to if they were singing above the surface.

When sound travels through something, it causes vibrations. For example, when someone plays a guitar, the instrument vibrates because of the sound waves traveling through it. If you put your hand on your throat while you're talking, you can feel your throat vibrating, too. Sound creates much less energy than other things, however, so it can't be used to create electricity or other forms of energy to power things.

CHAPTER TWELVE

ENERGY IN THE FUTURE

It takes a lot of energy in the form of fossil fuels to make things such as plastic water bottles and bags. When these plastics are melted, the energy in them could be used to heat homes and power cars, buses, planes, and ships. We don't have the **technology** to do this yet, but scientists are working on it!

Another source of energy we might use is called fusion. This is a process that powers the sun and stars. Fusion happens when the **nuclei** of two atoms of a certain element, such as hydrogen, fuse, or combine, into one new nucleus. The temperature needs to be very high for fusion to happen—up to 270 million°Fahrenheit (150 million°Celsius)! Before we can use fusion to power our world, scientists need to find a way to quickly, easily, and safely heat up the atoms.

Using discarded plastic bottles as fuel could solve more than one problem. It would save fossil fuels, it would stop bottles from polluting the land and ocean, and it would make more room for other types of trash at the garbage dump.

CHAPTER THIRTEEN

NUCLEAR ENERGY

Fusion produces nuclear energy, but we already use other ways of producing nuclear energy to create electricity. Reactors at nuclear power plants create energy by splitting the nuclei of atoms of the element uranium. This process is called fission, and it creates heat. Workers at a power plant use the heat to turn water into steam and use the steam to create electrical power. Nuclear power can be a good thing because it doesn't heat up the planet the way fossil fuels do.

However, there are some problems with nuclear power. If the process isn't done exactly right, it's very dangerous. Special machines at the power plant keep the fuel inside, but if the uranium gets too hot, it can melt those machines and spread dangerous **radiation** to the area around the power plant. Nuclear radiation is very bad for living things.

There are 60 nuclear power plants operating in the United States. Some people think this is too many. Others think it's not enough.

CHAPTER FOURTEEN

ENERGY AND CLIMATE CHANGE

Every day, people use a lot of energy. We need it to help us get shelter, food, medicine, transportation, entertainment, and more.

The issues surrounding what kinds of energy we use and where they come from are complicated. Humans have changed Earth because of how much energy we use. Fossil fuels are the biggest source of energy that humans use right now. However, every time we use them, we release gases into the **atmosphere** that trap heat here. This makes the whole planet hotter, which changes the Earth's **climate**. Because of global warming and climate change, a lot of plants and animals are dying. Weather conditions are becoming more extreme and dangerous. Scientists say that we need to stop using so many fossil fuels. It's time to look at other energy options!

GLOSSARY

atmosphere (AT-muh-sfeer) The whole mass of air that surrounds Earth.

biomass (BYE-oh-mass) Plant or animal matter used as a source of fuel.

chemical energy (KEH-mih-kuhl EH-nuhr-jee) Energy stored in chemical compounds and released when a chemical reaction happens.

climate (KLY-muht) The weather in an area over a long period of time.

fossil fuel (FAH-suhl FYOOL) A fuel—such as coal, oil, or natural gas—that is formed in the earth from dead plants or animals.

nucleus (NEW-clee-uhs) The central part of an atom, which is made up of smaller parts. The plural form of the word is "nuclei."

photosynthesis (foh-toh-SIHN-thuh-suhs) The process in which green plants use light to turn water and carbon dioxide into food.

radiation (ray-dee-AY-shuhn) Waves of energy.

renewable (rih-NEW-uh-buhl) Able to be renewed, or restored or replaced by natural means.

reservoir (REH-zuh-vwar) A man-made lake used for storing water.

technology (tek-NAH-luh-jee) A method that uses science to solve problems and the tools used to solve those problems.

turbine (TUHR-byn) An engine with blades that are caused to spin by pressure from water, steam, or air.

INDEX

PRIMARY SOURCE LIST

Page 7
Illustration of how 19th-century life was changed by steam power and electricity. Lithograph with modern watercolor. 1876. Everett Historical, held by Shutterstock Images.

Page 23
Olympic champion Shaun White competes in the men's snowboard half-pipe final at the 2018 Winter Olympics in Phoeinix Snow Park. Photograph. Leonard Zhukovsky. February 14, 2018.

Page 29
Nuclear reactor on the Hudson River, north of New York City. Photograph. Held by Shutterstock Images.

WEBSITES

Due to the changing nature of Internet links, PowerKids Press has developed an online list of websites related to the subject of this book. This site is updated regularly. Please use this link to access the list: www.powerkidslinks.com/SOPS/energy